1939

A YEAR IN THE RURAL DORSET LANDSCAPE

EDNA RICE

Illustrations
ROSIE RICE
Maps
SEAN RICE

RICHMOND HILL PRESS

First published in the United Kingdom in 2005
By Richmond Hill Press, 11 Barleyfields, Gillingham, Dorset SP8 4UN
Email: richillpress@hotmail.com

Copyright © Edna Rice 2005

Illustrations copyright © Rosie Rice 2005

Maps copyright © Sean Rice 2005

All rights reserved. No part of this publication may be reproduced, stored in a retrieval system, or transmitted, in any form or by any means, electronic, mechanical, photocopying, recording or otherwise, without the prior permission of the publisher and copyright holder.

British Library Cataloguing in Publication Data.
A catalogue record for this book is available from the British Library.

ISBN 0-9548000-1-X

Printed in Great Britain by Salisbury Printing Company Ltd., Salisbury.

Dedicated to my dear children –

Rosie, Sean and Kevin;

my grandchildren and great-grandchildren,

for all their love and kindness.

Acknowledgements

I would like to acknowledge everyone who so kindly helped with this little book. Rosie who worked so hard creating the illustrations, Sean for drawing the maps and giving advice and Kevin who always encourages me and lectures me on science!

I must also thank Dr Jack Skelton-Wallace and his wife Enid of Richmond Hill Press, without whose great help and encouragement this book would never have seen the light of day. For their enthusiasm and attention to even the smallest detail, I am very grateful.

I must also mention Liz Fricker for starting the ball rolling and writing the Foreword.

Foreword
by Liz Fricker of Gillingham
Freelance writer

I first met the author, Edna Rice, a few months ago. She was attending a meeting of the Gillingham WI when I was talking and singing to her group. As a freelance writer and community printer, I was talking about my work and, after the meeting, Edna spoke to me about her desire to publish a book.

When I received her letter, briefly describing her life and her attempts to get a book published, it was clear that this lady had something of interest to say. When I was at school, history was one of my worst subjects as we sat in the classroom with a teacher intoning lists of dates and facts to us which the pupils were supposed to write down and learn. How refreshing it was in later years to discover social history, from people who were actually there. They told of their personal experiences of a time in the past and were witness to the events which changed the world. In her letter, Edna said: 'I have now passed my 80th birthday and this will definitely be my last chance to get published.'

How could anyone fail to have been moved by Edna and her desire to get her book published? Thankfully, as a writer, I meet people who have achieved their dream of getting a book into print. Amongst the authors was Dr Jack Skelton-Wallace of Gillingham, who had already achieved this dream. Putting him in touch with Edna has, I think, been a source of inspiration to her and, more importantly, he knows the way through the book publishing process.

By the way, Edna also has two more 'strings to her bow' when it comes to producing an interesting and informative book. She writes poetry, as you will see, and has a daughter who is a gifted artist and has provided the drawings to illustrate this book. In this sad, mad, bad old world, it's great to meet someone who has taken the time to document past events from a personal point of view and give those of us who weren't there an insight into the history of World War II. Thank you, Edna, you have done a great job!

Liz Fricker

2005

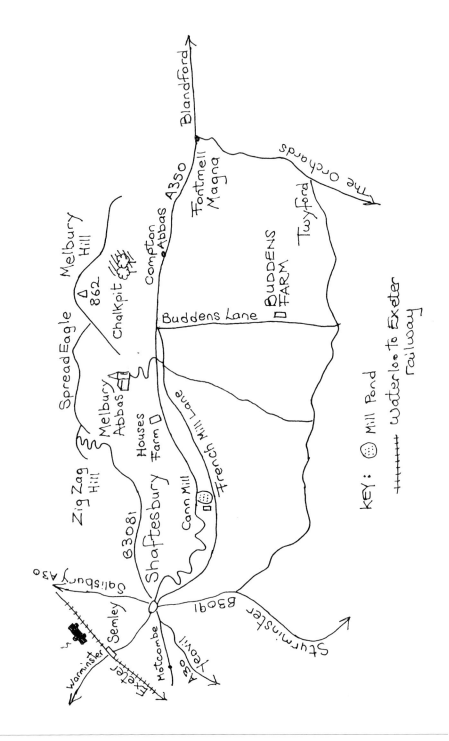

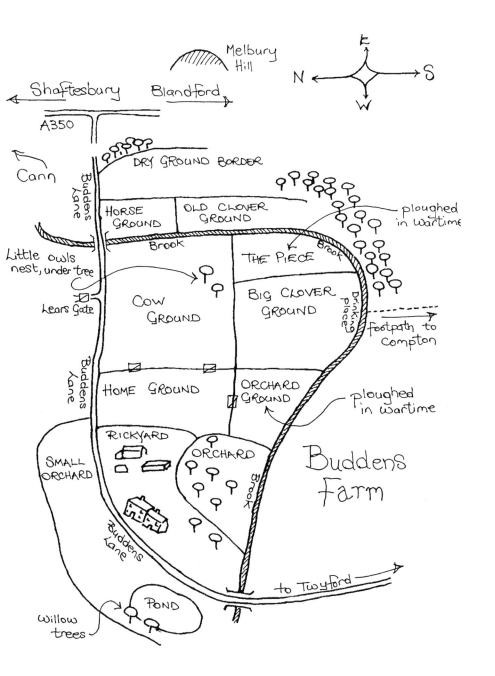

Introduction

The authoress of this superbly evocative book, Edna Rice, was born on 6 December 1924 at Houses Farm, her grandparents' home at Melbury Abbas, close by Shaftesbury, North Dorset. From birth until she married in 1951 her daily life was rooted and grounded in farming, specifically at Buddens Farm, her home and in the countryside in general.

Edna was educated at the local village school in Melbury Abbas, which still remains, and which she left at the age of fourteen to work with her parents. Looking back now on the days of her childhood and as a young woman, she has not one feeling of regret that life could have been better elsewhere. Reading her journal one gains the feel and taste of the sheer dogged hard work in running a farm; daily and seasonally, passing through the freezing, body-shrinking, fox-barking Winters, the spirit-nurturing, burgeoning and blooming Springs, the bakingly hot, bird-song-filled, scented Summers, and the darkening, reflective and introspective leaf-falling Autumns. From her early schooldays she possessed a deep passion for seeds and plants. Edna also developed an interest in painting, taking lessons in the medium of oil in Shaftesbury.

To write that Edna Rice is a very modest lady of simple tastes, somewhat self-effacing in fact, is not to suggest one trace of insult. On the contrary, as the great Greek historian Thucydides observed, 'The simple way of looking at things is so often the mark of a noble nature,' that distinguishing mark of nature is firmly implanted in Edna. Her style of writing is direct and uncomplicated. When one reads this very valuable social document it is Edna Rice speaking; enthusiastic, moving easily from the subjective to the objective and back again. One fascinating glimpse of her internal world is revealed by the fact that when she writes poetry, her work has to be penned out of doors, in the fields, her whole physical, mental and spiritual being surrounded, enveloped by the natural environment which was and is now, to a lesser degree, her entire life; no library author she. It is her simplicity of style and direct observation which elevates the mind of the reader onto a higher plane of thought.

Edna Rice treads the same deep furrow as many other farming and rural writers down the centuries: from the Greek poets, Theocritus and Hesiod; through Virgil, the supreme writer of the Gold Age of Augustan Latin poetry; England's own, the sublime John Clare; the Jesuit, Gerard Manley Hopkins, whilst, geographically much closer to home, Thomas Hardy and the tender, William Barnes.

Jack Skelton-Wallace

January

The New Moon

*I saw the new moon,
A silvery thread,
Over the ash tree,
When the day was dead.*

*The sky in the west,
Had an orange glow,
Lights in the houses,
Shone bright in a row.*

*The world sank to silence,
The frost shining white,
The moon in her heaven,
The queen of the night.*

January

New Year's Day seems a good day to start this memory of long ago; 1939, the year World War II started. It was a lovely sunny morning for the first of January, the land looked beautiful, the sun was trying to peep through the low clouds, making shadows and bright patterns on the patchwork of fields and trees running up to Melbury Hill, a landmark of 862 feet, where hundreds of years ago, in 1588, the beacon was lit to give warning of the Spanish Armada to the farming folk living around it, and even the ones all around the Blackmore Vale and beyond. In more recent years it was lit for celebrations of the Silver Jubilee of King George V (1935) and the Coronation of Queen Elizabeth II (1953).

The small copses were bare and quiet and the little brooks sparkled where the sunshine touched them. All was peaceful on this first day.

I had just left Melbury Abbas School at fourteen to help on the farm 'Buddens,' which had always been my home; my father was born there in 1898.

It was such a beautiful part of Dorset, enjoying some of the most productive land in England. The soil was heavy clay, but it grew a prodigious amount of grass and was like the poem about Barbara Frietzsche and Frederick Town, *Fair As The Garden Of The Lord* by John Greenleaf Whittier. In the summertime or from the top of Melbury Hill, in the words of the song, 'The valley lay smiling before them.'

The day started as any other, getting up about six thirty to get the milk ready for the milk lorry which came at eight o'clock. The sixteen cows were hand milked after getting their feed of hay and a dish of cake for the ones giving the most milk. They were tethered all the time, as the land was too wet for them to go out in the fields in the wintertime. After the milking was finished and all the dry stock had been fed it was breakfast time, always the same fare, home-cured bacon and eggs with fried bread, it was scrumptious. Sometimes we had some Force, which was like cornflakes, it was made with maize, and on the packet there was a picture of a funny man, with a queer kind of hair style sticking out from the back of his head, leaping over the fence, with the caption, 'High o'er the fence leaps Sunny Jim, Force is the food that raises him.'

At breakfast the programme for the day was usually discussed, but as the day was Sunday we just hurried through our various tasks to be ready to cycle off

(Mum and I) to Melbury Church for the eleven o'clock service. We had to be in good time as Mum was the organist. As it was still near Christmas we sang our favourite carols and there were a few more people in the congregation, as some of them were still at home for the holidays. Usually, except for holidays, weddings, funerals and harvest home there were only a handful of people there.

Opposite the church was the village school which I had so recently left. In the playground surrounding it were four large trees, horse chestnut and limes, what fun they were for childish games, and in the centre was a square bricked area where we did our PE and played such games as 'Puss in the Corner,' and 'The Farmer's in his Den.' We had a first hand view of all the goings on at the church and rectory. Those days were now over however. We cycled home from church, chatting to friends and neighbours on the way. In the afternoon I went for a walk with one of my Hayter friends from Twyford, which is a tiny hamlet adjoining our farm. They lived there in the most beautiful old thatched house, sadly, now knocked down by the council. We went to the top of Melbury Hill, making sure to be back just after four o'clock in good time for milking.

There was a visitor for tea, my grandmother on my father's side; she came from Compton Abbas in the pony and trap. This was the very end of the horse and trap era, though I don't think we even noticed its passing.

Sometimes Granny rode in the side chair of Dad's motorbike. Granny was a very special lady, she was a real tomboy, in spite of her ways of trying to bring me up properly. We all spoke with a pronounced Dorset accent, shades of William Barnes, and I might say, 'Look at thic oon there,' 'That one there, my dear,' said Granny. I'm afraid it did me little good.

Our day came to an end after sitting by the warm fire, and it was time to light the candle and away upstairs to sleep in the heaven of our warm feather bed; we were real feather bed farmers.

That was New Year's Day. From now on I am going to look back over the years, and take the month of January as a whole, as many days were almost repetitious of others, except for the weather. The cows took up more time as they had to go out in the yard to drink twice a day. They had to be fed about five times a day with hay and once in the mornings with mangolds. I loved to go into the stall first thing in the morning and smell that warm hayie breath. It was nearly always beautiful smelling hay they ate, as Dad was a canny farmer weatherwise, he very seldom made bad hay, and the fields had the natural meadow grass, with only natural fertiliser and no weed killers. I did not have much time to stand and sniff though because the cows would be hungry and they would all toss their heads and rattle their chains for the first meal of the day, which was placed in the forestall in front of them. Of course it was pitch dark on January mornings and we used a couple of hurricane lamps to light the stall. Oh how I would love to be able to go back just once and see and smell those things again! All our cows had names; Ruby, Pip, Dolly, Clover, Baby, Bluebell, Dinah, Orange and Lily were some of them. They all knew their own places in the stall, they never went in the wrong place. I loved to bury my head in their warm bodies and fill the bucket with creamy milk. They all had different characters, some bossy, some cranky, some placid, some nosey etc. just like people. They all had horns then, Dolly's were very long, but she was not one of the bossy ones, so did not use them much. We were lucky to have such a lovely modern cowshed, it was only built the year before; we got a few pence a gallon more for accredited milk if the stalls passed the test. We had to wash the whole place down every day, as we had lately got the water laid on this was possible. Before that the water was all drawn from the well in buckets and carried the hundred yards to the stall. We also kept a shorthorn bull, (most of our cattle were shorthorns) he lived in a house on his own, he went out for a drink every day and we always had a bowl of cake ready for him, so he would rush back in as quickly as possible. He was a very well behaved bull. Our bulls were always called Billy. We kept most of our heifer calves, but the bull calves were sent off to Sturminster Market when they were a few weeks old, no one sent off poor little day old calves as they sometimes do nowadays, which is a cruel practice. I couldn't bear to see any animal go off, I loved them all.

A lot of time was spent each day preparing food for the cows; as well as the hay they would each eat half a bushel basket of pulped mangolds. Dad always insisted that they must be meticulously cleaned with old knives before being pulped in the hand-turned pulper. This was pretty hard work, the mangolds looked like lovely white chips when they came out. If the hay was short, oat straw or second class hay, though only as a last resort, was put through the chaff cutter and mixed with the mangolds.

Somehow all these tasks were completed by about half past ten, as during the winter there was rabbiting to do. The rabbits were a real scourge, there were hundreds around some fields, the fields of corn were ruined by them, and there was often about ten yards bare around the hedges. Dad got the rabbiting from many farms around the area, and caught them by means of wires, nets, guns, dogs and any other way possible. A man came to collect the rabbits about twice a week, so we usually caught them for a couple of days before that. After we had been indoors for a hot drink, the ferrets were put in a bag, the dogs were rounded up and we were away with a hunk of bread and cheese to eat for our dinners, the dogs usually ended up eating most of it; they would look so imploringly at us. If Dad had wires down he would have been round earlier checking them, but sometimes the fox got there first; the poor rabbit could not escape.

One snowy day Dad had rabbiting taken near Ashmore, a distance of about five miles from home, the snow was too deep to bring the motorbike and he had wires set there, so we walked there by way of Melbury Beacon. We found about thirty rabbits. We paunched them which made them quite a bit lighter then we hurdled them, by making a small slit in the skinny part of the leg and passing the other leg through, then tied them into two bundles. We put them over our shoulders, half hanging in front and half behind. Then we walked home again over the tops of the hills through deep snow. Those rabbits were worth about four pence each in old money. Our lives were so different from most of the young people today, they would never do it now or even understand our lives, but I enjoyed it all.

There was one day a week when there was no rabbiting, that was Thursday, Shaftesbury Market day. Dad always went to it and there were always confloptions with him getting ready. First of all he got out his big cut-throat razor and gave it a good sharpening on the strap. After he had shaved, which was only done about twice weekly, he would put on his clean shirt, but he could never find the collar stud, the language began to get a bit thick until it was eventually found and after much fumbling got into the shirt. The shiny boots and gaiters were put on. There were papers to be found as all bills were paid and business done on market days. The bureau was turned upside down, paper going in all directions, but the missing ones would completely disappear. I used to keep out of the way as much as possible until he was finally gone up the lane in frenzy on his motorbike. I'm sure as soon as he arrived in Shaftesbury he went into the pub and was happy as a lark in a few minutes, he deserved his holiday. There was a sort of holiday atmosphere at home all day, although I was busy around the yard most of the time. Dad often bought a heifer and calf or anything else he thought was a bargain at the market, and he usually arrived home with a large piece of salt fish or a pig's chap. Sometimes some of the animals had to go to market, that was a sad day, which I do not wish to recall.

February

Starlings

The sun sinks in an orange sky,
The land lies drear and bare,
In, the wandering starlings fly,
    Starlings everywhere.

In the ghostlike trees they land,
Like black magnolia flowers,
Raising their throats in chorus grand,
    As the sunset lowers.

The wintry air is touched with frost,
The evening shadows fall,
The starlings seek their sheltered roost
    And peace is over all.

February

The days are getting a little longer now and we are looking for Arthur Street's Spring Day. Mr Street was a famous Wiltshire farmer author, who said February never went by without one spring day, and he was usually right. We were looking out for the first primroses, we knew all the spots where the earliest ones were to be found. Beyond Dryground border was a small steep corner of a field, facing south and sheltered from all other directions, there among the dead bracken and blackberry bushes were the loveliest early primroses. We posted boxes of them off to our city friends.

As we neared the spring the rabbiting lessened, but there was another job to be done – hedging. Dad tackled at least one big old hedge around the farm every winter. First of all the brambles were cut back with a slash-hook, and then most of the big stuff was cut, leaving enough to lay. Dad was very good at laying a hedge, he would finish up with a hedge so thick no animal could get through, and so neat and tidy too. We then had to sort out all the wood, the big stuff was sawn up for the fire, some was kept for fence posts, some for bean sticks, and the young hazel rods were kept for making spargads to peg down the straw when thatching the ricks. Then all the brushwood was made into faggots for lighting fires and boiling the copper. The faggots were tied with bonds made from twisted hazel sticks. The points of long thin sticks were stuck into the ground and then twisted until a loop was formed about four feet from the end. It was put around the faggot and the end was pushed through the loop, and it was bent over by the same twisting method. You needed very strong hands to twist the sticks. The few thorns and trumpery, as Dad always called rubbish, left over when everything else was gathered up, was then burnt. Nearly everything on the farm was made use of.

The method of transportation on the farm at that time was by the Overlands, two American motor cars of about 1920s vintage, which Dad bought after the pony died. In some ways Dad had very modern ideas, I never saw any of our neighbours with motor transport at that time on their farms. Of course, they could only be used when the land was dry enough, although we used to put chains on the wheels of the one which was converted into a lorry, and it would go most places. It hauled the wood, hay, mangolds, manure, everything in fact. The other one was used mostly at haymaking time for pushing the sweep. It finished its life providing spares for the converted lorry. We took nothing to the garage. When the cars broke down Dad was able to mend them. He took the engines completely to pieces and laid them out on the carthouse floor. I wondered how he knew how to put them together again, but he did. He never had any training in mechanics and, apart from the motorbike, no experience, he had always worked with horses and carts.

One day we were cutting the hedge by the brook in Orchard ground, it was very frosty with a bitter north east wind, I missed my footing on the bank and fell flat on my back in the icy water, the shock made me gasp for breath, but it was worse having to run across two fields to reach home in my soaking clothes, which froze as I ran. How glad I was to see a blazing fire, I thought I would never get warm again.

Dad kept a strict timetable throughout the day, in the morning the milk lorry arrived about eight o'clock, so we had no choice about that, but dinner was always at one o'clock, and out again sharp on the stroke of two. The preparation for afternoon milking started at four, and teatime was always at six o'clock, so we had quite a long evening; bedtime was ten o'clock. Lots of our neighbours milked their cows at any time during the evening. 'Howling about all night,' as Dad put it. It was strange what a mixture of old fashioned and modern behaviour Dad was.

We had a large garden at Buddens, it ran all around the house, and a few years previously had been quite a dump. There was a big yew tree by the kitchen door and some very old box bushes and lilacs by the gate. There were quite a few plum trees and a cherry, near the cherry was a lovely egg plum tree, which never failed to produce a heavy crop, under this was a gooseberry bush, a very sweet one, and in front of that was the grinding stone. I used to hate to hear the words, 'Tudge, come and turn the grinding stone.' It was on a small stand, with an iron handle, and it was constantly in use for sharpening hooks, scythes, hay knives etc. Our garden had to be built around all these things, but over the years it became lovely. At this time of the year we would find patches of snowdrops coming out, and the spears of daffodils appearing. Spring was really coming.

One of my greatest delights at this time of year was to see the enormous flocks of starlings, whole fields would be full of them, gradually collecting together about four o'clock in the evenings before flying off to their roosts in Motcombe Woods. They would sit in the bare trees and sing their hearts out, a million or so strong, it was really wonderful, and then at some mysterious signal they would all take flight together, the noise of their wings like a gun going off. Are they still there now I wonder, are their sleeping places still uncut? Sometimes when the weather was foggy the poor birds became disorientated, small flocks would fly wildly in all directions, hither and thither quite near the ground, making worried chirps. When darkness fell all was silent, I suppose they had to take what shelter they could find in the nearest bushes.

March

Rooks and Seagulls

Stormy weather,
Gulls arriving,
On the furrows,
Dropping, diving.

Rooks and seagulls,
Black and white,
Flashing wings,
Spread in flight .

Reeling, rising,
Dipping, falling,
Landing, fighting,
Feeding, calling.

Seagulls soaring,
Whirling, weaving,
Rooks more so,
Staying, leaving.

Cawing, squawking,
Madly rushing,
On the furrows,
Roughly pushing.

Birds of the wild,
And stormy seas,
Sharing, with birds,
Of fields and trees.

March

If March comes in like a lion it goes out like a lamb, or vice versa. So we always believed, and were glad when it came in like a lion and got it over with. Another of our sayings was, 'Fogs in March, frosts in May.' I often made a note of the date on foggy days, and sure enough when May came and the potatoes were well up, the frost would come on those mornings. The routine was much the same as before, as it would be until the cows were turned out. At least the days were growing longer, 'As the days lengthen the cold strengthens,' we said.

The thirteenth of March was an important day for me, although I did not realise it at the time. It was the day foxy came. When I went in for dinner Mum said, 'Henry Ridout has brought you a fox, he says it won't live, but he thought you would like it.' He knew I was fond of animals. The vixen had given birth to the cubs in his garden hedge and they had promptly been basked out by the dogs. Wuffie (where did that name come from?) was the only survivor, he was a sorry sight, small, blind and helpless, something like a new born kitten, hardly any fur, except for three white hairs on the end of his tail. We put him in a box of hay, with a hot water bottle and placed it near the fire. We did not know what to feed him on, so we mixed milk, sugar and water and fed him a teaspoon at a time, even getting up during the night for the first few weeks. He began to thrive for a couple of weeks, we knew when to feed him by his plaintive little cries. His fur began to grow and his little eyes opened, but then trouble started, we just could not keep him dry, as he was a dog fox and had no mother to lick him. In spite of our washing and powdering, just like a real baby, he just got wetter and wetter and eventually his little back legs became paralysed, he dragged himself about the kitchen on his front legs, and cried incessantly. I was nearly frantic by that time, we had all become so fond of him and the constant crying was terrible to hear. He probably needed a change of diet, we may have given him the wrong food. As luck would have it I managed to catch some mice in the rickyard, and gave these to Wuffie, he began to recover straight away, and was soon able to run about and play like any other little fox cub. At night times and when there was no one indoors he stayed in the chicken run outside, but when we were indoors Wuffie was in too.

Sometimes on Monday mornings I would be commandeered by my mother to help with the weekly washing, which was very different from the methods used today. The big copper in the back kitchen was filled with water fetched from the well, I suppose it held about ten gallons, and then the fire was lit underneath it, always with faggots and wood from the hedging. When it was hot enough it was put in two wash tubs, a large one for the white wash and a smaller one for the coloureds. The white wash was bed linen, towels, tablecloths, shirts, milking coats and hats, in fact nearly everything except woollies. Plenty of soda was put in the water and a bar of soap in the galvanised dish which fitted on the tub. There was a wash board made of wood and corrugated iron, on which the clothes were soaped and rubbed. At last it was put in the copper and given a good boil for about twenty minutes. It was then taken out and rinsed twice, the second rinse got a good swish around with a blue bag and plenty of starch was added. After that it was put through the big mangle with wooden rollers. The coloureds were also washed and mangled the same way and all was hung out in the fresh air to dry. The flagstone floors were then washed with the water left in the copper. The clothes we washed were mostly cotton or linen. We used to buy unbleached sheets and after a few boilings they became very white. The pillow cases were all handmade, as were the night-dresses, every year Mum bought a large piece of winceyette for these. It was quite unheard of to buy ready made things. The feather beds were home made too, the feathers were saved from any poultry cooked for dinner, all the big feathers were disposed of, the others were all picked over and the hard bits cut off with scissors. Then a large bag was made by stitching a few sheets of newspaper together and the feathers were put in this, and placed in a slow oven until it felt crisp and hot and the paper became brown, this got rid of any creepy crawlies and germs. When enough had been saved we made a large tick with stripy ticking material. Anyone who has never slept in a feather bed does not know what they have missed. The beds are so warm, you just sink into them and there are no spaces for cold air all around. They have to get a good shaking every day, many the times I would help Mum by taking a corner at one end, and she the opposite one and we would have a good shake and often a good laugh as well. Of course, the quilts were handmade too; Granny made most of these with patchwork.

I can remember seeing my grandmother washing and ironing smocks, which my grandfather used to wear on the farm, but I cannot actually remember seeing him wearing them. When the clothes were dry they had to be ironed. We heated the irons on the fire or the oil stove, there were always two so one could be heating up while the other was in use, and they had to have a good rub as sometimes they got quite black from the smoke. After the washing we always had a dinner of cold meat and bubble and squeak (as the fried potato and cabbage were called), left over from the day before. There was a big open fireplace in the kitchen, beside the copper, over which hung a big iron pot filled with potatoes etc. for the hens and pigs. I was always being admonished in a semi serious, jokey sort of way with the rhyme,

> *Dearly beloved brethren, isn't it a sin,*
> *When you peel potatoes, to throw away the skin,*
> *For the skin feeds the pigs, and the pigs feed you,*
> *Dearly beloved brethren, isn't that true?*

Quite a lot of the farm jobs were done by the fire on bad days, as the kitchen was very large. Sometimes in the summer times we had our meals there, as there was a paraffin cooking stove in the corner, with two burners and an oven, which could be stood on the top. Most people used paraffin stoves in the summer as there was not much electricity then.

Sometimes the weather would be good, and we got a warm day, Dad would come in with glee and say, 'I jis seen a girt dumbledoor,' meaning a large bumble bee. He was always looking for signs of spring, and listening for the first chiff-chaff. They usually came about the last week of March. He used the Dorset names for the birds, *Bobby* for robin, *Cutty* for wren, *Polly washdish* for pied wagtail, *Stone thrush* for missal thrush, *Screech owl* for little owl, *Longtailed chattermag* for longtailed tit, etc. The animals had names too, rabbits were *conies*, the fox *Reynard*, and the moles were *wants*. He would go through the wicket (gate) and up the lane to the turnpike (main road), where he might meet some diddies (didecoys or gypsies). If he was surprised by something he would say, 'Well, sup me bob!'

April

Swallows

Welcome to you swallows,
In your coats of blue,
Welcome to your coming,
To the homes you knew.

Welcome to the blue skies,
Welcome to the streams,
Welcome to you darting,
In the bright sunbeams.

Welcome to your cosy nests,
On the old barn wall,
Welcome to your twittering,
O'er the big cow stall.

Welcome to your singing,
Songs of warmer climes,
Welcome to you swallows,
In the summer times.

April

April fools day, Dad would be trying to catch us out, always something about animals. We had a view of several fields from the kitchen window, and Dad used to say, 'Look at Reynard there,' or, 'That's a funny coloured rabbit,' he loved to see us crane our necks to see. It was getting nearer summer, and the cows knew it too, they would be smelling the air and looking longingly over the homeground gate, but they had to wait as there would be plenty of bad weather to come, we had blackthorn winter in front of us, when the east wind would blow up for weeks and burn up all the early grass. That was the time when a lot of poor animals were nearly starved, as the inefficient farmers ran out of hay and turned the poor animals out onto grass which did not grow, our cows were never like that, Dad nearly always had a rick left over, in case of a bad season to come and the old saying was true again, 'March will search and April try, May will tell if they live or die.'

Between the third and the sixth of the month the first swallows came, they came back to their old homes in the barn; how I loved to see them and listen to their chirps. The martins came too, they made their nests under the eaves and by opening the doors of the loft above the stable we could look right into them. The lofts were used, until that time, for hay for horses and cattle. The only doors were high up ones where the wagons could stop outside and the fodder was thrown up by prongs. Our only way into the lofts was by climbing up on the manger in the stable and through the hay racks, then pulling ourselves through a hole in the ceiling. The cattle cake was kept up there and the cake crusher. Many hours Mum and I spent cracking the big sheets of cake, she had to climb up the same way, and down again, not so easy in skirts. The farmers' wives very seldom wore trousers then, it wasn't the done thing. I sometimes wore a pair of men's dungarees or boiler suits.

The ninth of April was Easter day, everyone wore their best coats and hats. At that time the halo hats were all the rage and nearly all the ladies wore them. Mum and I were suitably attired in our best dresses etc. and then as every Sunday, rain or shine, we were away on our bicycles to church at Melbury Abbas. We had to be in good time as Mum was organist. She had played there ever since she was a young girl but she retired temporarily when she got married and they presented her with a sewing machine. There was no electric blower for the organ then, and someone had to blow it.* I was always hoping it wouldn't be me. I often blew it for choir practice, which was usually held one evening during the week. We usually sang an anthem at Easter as well as the hymns. Melbury is a lovely church, it is set below Melbury Beacon in beautiful countryside. It was built in 1851 on a former church site, by Sir Richard Glyn. It has a tower at the south west end which houses five bells, they really were a joy to hear. The tenor bell weighs over seven hundredweight. The bell ringers were very good at their jobs; they usually rang for half an hour before services. I was always fascinated by the tower and often went up the stairs to see the bells. One day I was allowed to go up to the top of the tower and look out over the parapet, the view was breathtaking. I took photos with my old Kodak camera, it was new then, it cost 12/6. (62.5 pence)

As April went on there were many jobs to do on the farm. Fencing was one of the most important to be done before the cows went out, all the rotten fence posts had to be replaced and barbed wire put up. This would not have been allowed twenty years before when the farms were the property of Sir Richard Glyn. Our cows were never allowed to stray.

The ground had to be prepared for potatoes to be planted and mangolds to be sown. Another job was stone picking. The manure which was put out on the mowing fields always contained stones, which were brushed up from the yard; we had to pick them all up in buckets, systematically covering the ground as we went. It was a pleasure on a nice day, as the grass was growing and there was a lovely fresh scent from it, and the daisies began to appear and the swallows would be flying overhead and the cuckoo calling any time after the sixteenth. All the birds were thinking about nesting. Ever since I was small I had collected birds eggs, never more than one from a nest, these were all carefully laid on cotton wool in a specially sectioned box, which my Uncle Reg had made for me. We went to great lengths to get them, Dad climbed up into the tall trees to get eggs from rooks, crows, sparrow-hawks, etc. for me. We found many ground nesting ones too, the peewits and skylarks were very hard to find, the parent birds went to great lengths to put us off the scent. The peewits feigned broken wings and fluttered along the ground, they would never fly straight up from the nest. We loved to find them, but we only took the egg if we had none in the collection.

The ground was drying out towards the end of the month, the spuds were planted and the mangolds were sown in well manured ground. The chain harrowing had to be finished before the grass grew too long. Things were

---

* blow it = pump the organ bellows by hand.

beginning to get busy. We were looking out for a perfect spring day to let the cows out. It was often May before we would chance it, but when the day came it was marvellous. We threw open the gate into the homeground, and away went all the cows, throwing their hind legs into the air, galloping and leaping in sheer delight, and trying to eat tasty morsels of grass at the same time. We were just as happy too, no more cleaning out wheel barrows of manure every day, no more hay hauling or mangold pulping, except for a few for Billy and a few heifer calves which were not ready to go out. It was really springtime! We often got a really hot day and no one had any energy, everyone said, 'Lawrence is about.' I never found out who Lawrence was or if it was only in our locality it was said.

May

May

Cast ne'er a clout 'till May is out,
The May is out, we'll cast our clout.
The cuckoos calling all the day;
Cuckoos always call in May.

Cuckoo flowers and Queen Anne's lace,
Make such a show about the place,
But the very best of all,
Are lilacs blooming near the wall.

May

May was such a lovely month on the farm, the cows are OUT! We are free at last. The countryside is a picture with flowers – the woods are blue with bluebells, the lanes are arched with may blossom, the banks are covered with primroses, campions, cow parsley and stitchwort, the fields are golden, first with dandelions and then buttercups and there are marsh marigolds by the brook.

The cows quickly eat off the early grass, it does not grow against them until about May 12th. About that date was the time for St James' common to open. Dad usually bought a few grazing leases, one for each cow, a horse is supposed to eat as much as two cows, so they had to have two leases. The common is situated just to the south of Shaftesbury, and about one and a half miles from Buddens. There were gates across the roads at the end of every road leading out from the common. We drove a few heifers up there, we only let them stay there a few weeks as they sometimes strayed when the grass became scarce. Most people were pretty good to close the gates after them. Quite a few people owned cars and a journey to town and back would involve getting in and out of the car eight times to open and close the gates. Of course there were always a few who didn't care and the animals could wander a long way, or even up into St James', and as the common covered large distances, it could take all day to find cattle.

The mangold seeds were all well up and waiting to be hoed, this was done as soon as they could be easily seen, and then again when they were singled, sometimes we had to transplant a few to empty spaces as we went along. When they were small the mangold and swede seedlings were often attacked by fly so we used to drag sacks soaked in paraffin over them. In spite of everything we usually had a good crop.

All this time little Wuffie was growing bigger, but he never lost his nervousness and he was always looking for ways to escape. I would have let him go rather than see him unhappy, but it would not be fair for a half tame fox roaming around the farms, he wouldn't stand a chance and would be in danger of being shot, as every

farm at that time had plenty of free range poultry. I worried about him a lot, and then one day the problem solved itself. He managed to push up the lid of his run and away he went, unseen by anyone. We were all on the lookout for him, and we often saw him by the thick hedge near the brook, but he would not come up to us, that bit of nervousness would not let him.

After four days we were becoming worried about him as we could see he was getting thinner and hungrier all the time, he was no good at all at looking after himself, so Dad decided there was nothing for it but we must catch him, not an easy thing to do with the thick hedges now all covered in foliage. He thought the best way would be to put down a wire, but it would have to be watched the whole time so Wuffie would not come to any harm in it. I spent most of my time down in the brook on watch, it was not very long before he was captured. Talk about a reformed character, it certainly taught Wuffie a few lessons, he put his little front legs around my neck and clung on for dear life all the way home. I thought he would never stop eating, and he was so contented and happy, he never tried to escape again.

We made him a nice home in one of the pigsties in the yard, which had an outside run to it, though we did tie him on, as none of the hens would be safe. He did get an odd one though, he would lie just inside the door and watch until a hen came along and he would let it pass him, and made no movement at all until it was well inside the door, and then there was no way it could escape. He got away with it with Dad too, if a wild fox took a hen it was hunted down unmercifully but there was not a word said when Wuffie committed the crime. We used to bring Wuffie indoors with us every mealtime, or any time we were in the house. He was completely spoilt, he would take the best chair in the house and be fed on the best food and he loved it; he knew every word we said. He did not like going for walks, he seemed frightened of everything he met, and I always ended up carrying him. One day I got a photographer to come and take his portrait, it was a very formal pose. I took a few black and white photos, there were no coloured films then and we had no idea of taking things naturally, everything was especially posed. Wuffie made great friends with the dogs and cats; they played games together. One day I found a little abandoned hedgehog baby and took it indoors in a box, Wuffie was very interested in it but he didn't like the prickles and would jump back, although they were only soft at that stage. I called it John James McBristles after a character in one of my old books, but when it grew up it wandered away.

We often had baby animals indoors, baby pigs mostly, they were often the weakly ones. We had to watch the sows when they were farrowing in case they lay on the little ones. We had a railing all around the pen about a foot from the ground

and about a foot out from the side that the little pigs could dodge in behind out of danger. In the summertime the pigs went outside quite a lot, the rickyard had iron railings which the pigs could not get through, so it was an ideal place for them. Sometimes, especially when there were acorns, they had the run of the whole farm and almost kept themselves in food. Of course they had to be well rung or they would have rooted everything out of the ground.

The Sunday evening walks were lovely in May, Dad and Mum and anyone else who was visiting went for a ramble around surrounding fields and often ended up on Melbury Hill. The loveliest sight of all was to watch the fox cubs at play. One of their favourite places was Dryground Border, which was a small wood at the top of sloping fields rising up from the brook; the wood was blue with bluebells, many of which were flattened by the rolling antics of the little foxes. They were nearly three months old, and there were usually about four of them. We had to be silent and motionless not to frighten them back into their den. Sometimes the mother fox joined them and sat and watched the games. We saw endless birds' nests; especial interest was in the larger birds, the sparrowhawks, buzzards, owls, magpies, jays, etc. These were mostly in the larger trees bordering the brook, but after we passed Dryground Border we came out onto the higher arable fields, and then we found the peewits nests, and would stop to watch their frenzied behaviour to put us off finding their nests. We went on even higher until we reached the bottom of Melbury Hill. This was the home of the skylarks. They sang all the time as we made our way up the steep sides of the hill. It was a beautiful sight to see the millions of cowslips and of course we loved the scent of them, which is like no other. There were many anthills, those queer mounds, some of them nearly touching each other, about a foot high and maybe two feet across. We kept mostly to the well worn path going up. When we neared the top there were gorse bushes in full bloom which were inhabited by rabbits. There were hares running on the hill too. The view from the top was wonderful; the Blackmore Vale was spread out before us to the South West and in the opposite direction was Wingreen. On a very clear day the needles of the Isle of White could be discerned. We often met up there; it was a favourite walk for local people on Sunday nights before the television came. We found some place to sit down on the springy turf but it was difficult to find a place where there were no prickly thistles; they grew so close to the ground you never knew they were there until you sat on one! There were lovely butterflies too, black and red spotted ones and delicate small blue ones; are they still there I wonder? I haven't seen them for many years. There were orchids too – 'goosey ganders' we used to call them. They were followed by harebells, those delicate waving bells of blue. I remember watching an air show from up there, which was taking place in the showground in Shaftesbury. The little planes were looping the loop and doing other aerobatics. They also took passengers up in the little biplanes.

Every year we picked dandelion flowers for wine, just the heads were pressed down in a quart jug to be measured, it was delicious to drink in the wintertime, along with elderflower, elderberry, potato, in fact nearly anything, even cowslip flowers, there was such a profusion of them at the time. We sometimes found clumps of oxslips growing along the hedges, which were a cross between the cowslips and primroses; we often planted them in the garden.

Another great joy in May was the nightingales' song, they sang night and day, but at night they had no competition from the other birds and were wonderful to hear; I loved that high pitched whistle at the beginning of their trills.

June

The Dew Is On The Roses

The dew is on the roses,
As every bud unfurls,
Each bright silken petal,
Encrusted is with pearls.

On the little wild rose,
With its sweet flushed face,
Dewdrops sit as regally,
As those of grander place.

June

One thing that was always uppermost in our minds when June came in – haymaking! It was the last year when the big cart horses belonging to my Uncle Joe were used to cut the hay. The day before they were to arrive the swathe around the hedge was cut with the scythe and raked back out of the way of the mower, there was never a blade of grass or hay wasted. The mower was in the field before seven o'clock. The carter who lived in Shaftesbury must have started out from his home at about five o'clock. He cycled down to Houses Farm about a couple of miles, and then harnessed his two big horses and hitched them into the mower and drove them another couple of miles to our house. They always brought nosebags of oats, and when the carter stopped for his big hunk of bread and cheese, washed down with some of Dad's cider, they would have their nosebags. By

the time we had finished the milking quite a lot of the field was done. Dad kept the knives of the mower sharpened for them, they changed them quite often. There was always hay to be raked back from the awkward corners. We usually mowed about four fields altogether. It was lovely in those hayfields early in the morning when the dew made the grass easier to cut. The swathes were full of scented flowers, the honeysuckle and roses lined the hedges, and the meadowsweet was beginning to come into flower.

The air was heavy with scent and pollen from the mown grass; the trees were thick and heavy too. Bees added their sound to the twittering of the birds, many of whom still had babies to feed. The cuckoo still gave an odd 'cuckoo,' though it was often 'cuc-cuckoo' now. The swallows were flying high when the weather was good and we were hoping it would hold until the hay was finished. After two or three days the hay was ready to be turned; we always hurried up to get out into the field as soon as the dew was gone. Mum and I wore our sun bonnets; we had no horse so it was all turned by hand with prongs. By the time we had it turned it was time to start at the beginning again and rake it into wheels, then after milking was done it was put into small heaps which we called pooks. (In some parts of the country they were called cocks.) Next day weather permitting, it was all shook out again to dry. Sometimes we were aided in the raking by our neighbour's mule, which pulled the horserake.

At some time previous to this, the staddle for the rick had to be prepared. The size would be according to the heaviness of the crop, and it had to be pretty accurate as there was no way of changing it later on. Some large boughs of trees were placed around the outside, and then others were put in to fill the centre of the oblong shape, with brushwood on top of that, followed by some of the most inferior hay from around the hedges. The two Overlands came into their own then with the haysweeps fitted on front, they flew along the wheels of hay and then deposited a huge mass in front of where the rick was being built. Dad and I drove the haysweeps, we sometimes pushed in quite a few loads (hoping it would not rain on them) and then with our neighbour Mr Kift building the rick, we threw it up with prongs, using longer handled ones as the rick grew higher. If we needed more help I went on the rick with Mr Kift. We would help him with his hay when the time came. When most of the hay was on the rick I spent many hours raking the fields with the mule and horse rake. It was lovely in the hayfields when Mum appeared with bottles of hot tea, or with lemonade made with little bottles of crystals. In the evening when we were finishing up for the day we sat in the hay, with whatever helpers we had, and Dad brought out the cider jar, and usually bread and cheese was shared around; those were happy times.

People were very good to help in the hayfields in those days, they knew how precious the hay was, it was a matter of life or death to the cattle and the farmers' livelihood, and they also enjoyed the cider! When the ricks were finished that was not the end of it, they had to be tucked all around, all the loose hay was pulled from the sides, until they were a perfect shape and would keep out the rain, they were smaller at the bottom for this purpose. When all was settled down it was thatched. We usually grew some rye for this purpose, cutting it green and tying it into sheathes by hand when it was dry. We had the spars made in readiness to peg it down. After that the rick had to be fenced around, only then could the cows go in the field to eat the after grass. Most of the month of June was taken up with haymaking as we were helping Mr Kift with his hay as well. This worked out very well, as neither of us had enough help to manage on our own.

All the other jobs on the farm had to wait while haymaking was in progress, it usually coincided with many other jobs; the mangolds and spuds were not being hoed and the weeds were growing apace. Time also had to be found to put in some swede seeds. The fruit in the garden was ripening but there was no time to deal with it; anyway we were too tired after our hard labours. We never did any but really necessary work on Sundays. I never knew Dad touch the hay on a Sunday, no matter what the weather. Although he was not a church goer, he was a very good living man; he always said his prayers at night. His God was in the farm and all the natural world around him. He usually had some of his cronies to see him on Sunday mornings when they read the racing in the Sunday paper, drank cider and discussed the crops and the neighbours. He used to lie down on the bed Sunday afternoons, as did many of the people at that time. Mum and I loved to sit in the garden. She made a hammock with two bamboo canes and some thick canvas and we used to hang it between the plum trees, which were festooned with Seven Sister roses; we took it in turns to lie in it, surrounded by the wreaths of roses.

We usually had some chickens hatching out at this time of the year. We tried to get the duck and goose eggs hatched out earlier because if the thunder came they might get addled. Dad said if the duck eggs were hatched after the first of July the ducklings would get cramp, we used to soak their eggs in water every day to soften the shells. As soon as they were big enough they ran about the farmyard with all the other poultry; it was all free range then. When the ducks grew bigger we had a lot of bother getting them home at night. After they discovered the brook away they went, regardless of time, and many of them made a tasty meal for a fox.

The calves were usually turned out about this time of year and they went in the orchard for a while. When we had dragged them unwillingly across the yard and in through the gate, they went completely berserk, never having been out of their small, darkish homes. To be turned out into the big world was more than they could cope with; they galloped at great speed, not knowing when to stop – into hedges, fences or anything which got in their way. However, they soon settled down when they discovered the grass was nice to eat.

July

A Wet July

The cows all stand beneath the hedge,
Their feet are sinking in the mud,
Their backs are hunched, their coats are wet,
They miserably chew the cud.

They walk away, they try to graze,
Rivulets running down each face,
A lake has formed inside the gate,
What's happened to this summer place!

Sweeping in quickly from the West,
The clouds are heavy, dark and grey,
The fields are all a swampy mess,
There is no chance to make the hay.

Raindrops in puddles making rings,
Swallows fly low around the trees,
Foxgloves droop their heavy heads,
The honeysuckle has no bees.

Maybe next week it will be better,
Perhaps we'll get a change of moon,
Maybe it will not be wetter,
England's summer is coming soon?

July

The summer is passing though we haven't noticed much difference in the day lengths yet. Thunder clouds loom up out of the blue, and the weather is often broken up for a few weeks. If the hay is finished we don't mind; the rain will do good and bring on the grass and crops.

About this time Dad bought the tractor. There had been rumours of war for ages and Dad, being foresighted as usual, decided it was necessary to be prepared for any contingency. There were very few tractors around then, only some of the bigger farms had them, although it was not long before a lot of farmers followed suit. Ours was a Fordson, a big orange coloured one, with iron wheels which had large cleats attached to them. Of course it was no use without machinery to go with it, so we got a few of the most necessary things: a plough, mowing machine and trailer, first. The carts started disappearing to make room in the carthouse for the new things. At that time we had a milk cart, a governess car, a dungpot and a nice wagon, which was later converted to be pulled by the tractor, for hauling sheaths of corn; it was painted blue and had large racks to fit on the front and back. Like most wagons it had the farmer's name and address painted on the side. There was also a lot of harness; some of it hung in the stable for many years. The tractor made life a lot easier as it could go anywhere in any weather.

It was lovely when the hay was finished; there was more time for everything. We got the fruit picked and made into jam, or preserved in Kilner jars to keep us going in pies all the winter. On nice summer evenings we set off on our bikes with buckets or baskets to pick raspberries on Fontmell Hollow. The lanes were a picture with summer flowers: marguerites, meadowsweet, bedstraw, bellflowers (which were called Canterbury bells), foxgloves etc. and it was beautiful cycling along by the river at Fontmell, where the swans swam with their young ones. It was a very steep climb up Fontmell Hollow. Right on the top of the hill the raspberries grew; there were huge clumps of them, some red and some yellow. They were very big and juicy and we stayed until our baskets were full. Lots of people went to pick them but there

were plenty for all. Riding home with them on the handlebars, the juice often ran out and trickled down our bare legs. Our cherry tree was quite big and every year produced a good crop, they were delicious for tarts; we bottled a lot of them.

There was often an artist staying in Twyford in the summertime, his name was Mr Spencer, I wish I could remember which Spencer he was, I know he was an RA he sometimes painted pictures around the farm. This time he spent some days at the bottom of homeground, sitting on a little stool, and painting the oak tree which leaned over the orchard ground gate. It was not a particularly pretty tree, a bit lopsided, but I suppose he had his reasons. I am always looking out for that picture.[†]

I loved fetching in the cows for milking, walking along well worn paths, with the buttercups swishing my bare legs. Sometimes they came when they were called. Everyone called in cows at milking time and you could hear them calling all around the district; it's many years since I heard anyone do that. We used to talk to the animals a lot then. When the war was on, one of our neighbours had a cockney evacuee boy, and he was told to hitch the horse into the cart. He had a lot of trouble with it and the farmer said to him, 'Why don't you speak to the horse?', 'What abaht?', said the evacuee.

All the fields on the farms had names, but ours were not very exciting: Homeground, Cowground, Cloverground, Orchardground, Horseground and a field called 'The Piece.' Some of our neighbours' fields were Big Bere, Little Bere, Sparvlings, Kernels, Brockley, Woodhouse Mead, Inside Mead, Dryground, Coltsground, and Roadground. Roadground belonged to Mr Kift; it was up Buddens Lane, joining the end of our land. There was a gate across the road at each end of it which had to be opened and shut every time we went up or down the lane as the cows were nearly always there.

One of the jobs which came up every year was thistle cutting; I spent hours going around the fields with a scythe. 'Cut a thistle in May, work thrown away, cut a thistle in June, come again soon, cut a thistle in July, it will surely die.' This was very true. Although I was quite good at mowing with a scythe, I was no good at all at sharpening it, so Dad usually gave it a good edge in the morning and it got progressively worse, blunter, and made hard work of it by the end of the day, in spite of all my efforts with the whetstone. There was always the chirping sounds of grasshoppers for company, I haven't heard one for many years, they seem to be gone along with the glow worms and the nightingales. There were quite a lot of wild orchids in the fields, mostly

---

[†] This Mr Spencer would have been Gilbert Spencer RA (1892-1979) the one year younger brother of the more famous Stanley Spencer RA. Gilbert spent a number of years, during the summer months, painting in this area. Of particular relevance is his *Blackmore Vale From Compton Abbas*, painted in 1942 and now in the Tate collection.

goosey ganders, but sometimes we found the bee orchids which were just like bumble bees, and more rarely the butterfly orchids.

In the garden the flowers ran riot – they set off the old stone house so beautifully; on the sunny side it was covered with roses, at the end wall by jasmine and on the north-west side, at the bottom of the wall, were ferns and peonies. Outside the kitchen door at the back was another lovely part of the house, the veranda. I loved it so much; part of it was open sided, with railway sleepers to support the roof, these were soon covered with roses and clematis. The deck chairs were put there for a few precious minutes of the dinner hour. A lean-to greenhouse had recently been joined onto the lower rendered end, and we could get to that from the house without going outdoors. Wuffie the fox used to like going in there too, it was so near the back door. Of course, the greenhouse soon became planted up; covering the back wall was a large blue plumbago and peach tree. It was laden with luscious fruit every year. There were tubs of lilies, agapanthus and arums, a few tomatoes, and Dad's favourite celery was brought on in there as well as bedding plants. It was also useful for baby chicks and ducks; Wuffie was kept out then! The

veranda covered the wall right up to the back kitchen door, this part included an old stone sink and pump, and we did use this at one time; we had to pour quite a lot of water down inside of it to plim up the leather washer, while pumping the long handle as fast as we could. By the time we could get any water out we were quite exhausted. At least we did not have to worry about drinking water for the cows in the summertime; they could drink from the little brooks, which ran through parts of every field on the farm.

August

August

*I see August flowers,
As up the lane I walk,
Flowers of every colour,
A bobbing on each stalk.*

*Foxglove, clover, buttercup,
Trefoil, meadowsweet,
Honeysuckle, willowherb,
Bramble and marguerite.*

August

The summer is passing, and the cornfields are beginning to turn colour to harvest once again. We had no corn in all my years at Buddens until then, but this would soon change when the war came. The corn was mostly on the higher ground. My grandfather at Compton, and Uncle Joe at Melbury, both grew quite a lot of corn and we were often in the fields at the end of the reaping when the centre of the standing corn became smaller and smaller. The binder would stop work just before the end, and the poor rabbits were trapped in the middle. There were always plenty of men with dogs and guns and the destruction was terrible. Our old dog Wendy ran so fast and caught so many rabbits, she became quite exhausted, the days were very hot too, and poor Wendy lay on the ground, unable to move, we thought she would not recover, but after a while she rallied round. I loved the scent of the corn at harvest time.

We often did some manure hauling while the ground was so hard and dry; we did it in spare time during the summer. It was all wheeled into an enormous heap, just through the yard gate into homeground, during the winter. While we were clearing this we nearly always found clutches of grass snake eggs, they were all together in heaps, with white leathery skins. We used the blue Overland to haul the manure as it was easy to load, with no sides, and quick to run around the fields. We usually left the manure in large heaps until the winter. The ground, all around the gate and manure heap, was covered with camomile, which gave off a lovely scent when we walked on it in the hot sunshine.

The cow stall had to have a good clean up too, every inch was scrubbed and the windows cleaned. The top part of the walls were whitewashed, as were the inside walls of the other farm buildings with home-made whitewash of lime and water.

The stable was used as loose-boxes for cows, many a time I was sitting in there milking a cow and I saw little bright eyes looking at me from the holes at the bottom of the old wooden partition, at first I wondered whatever they belonged to, but then something darted in an out, and it was a little lizard. I became quite fond of those little lizards. Dad used to get great pleasure watching the buzzards circling around the farm. There were the parent birds and two or three young ones. They always kept fairly close together, and were a lovely sight with their big wide wings soaring above the trees. There was also a large bat population, lots of them spent their days hanging upside down in the lofts, sometimes little ones fell down into the cow stalls and I always picked them up and took them back to their homes, and sometimes got bitten for my pains. It was lovely to see them diving around the garden in the twilight.

The garden was beautiful all the summer and took a lot of our spare time to keep it nice. Madonna lilies grew along the wall one side of the back door, they were in  their element there, about four feet high; such pure white petals and the air was full of scent. If you went to smell them your face would be covered with yellow pollen. The wall the other side of the door was covered with white jasmine so the scent of the two combined was really marvellous, these flowers were probably planted by my grandmother in the 1890s, or maybe many years before that; the fruit trees were most likely planted about the same time and the box bushes many years before that. We planted a lot of new plants and shrubs and we made an herbaceous border along the high wall at the back of the carthouse. There was a large flowered orange blossom at the back, also, on the wall, was nailed a pigeon house (two-storied) for the tame pigeons. We planted a large selection of plants; some of my favourites were red bergamot, phlox, lobelia cardinalis, red and yellow geums, and campanulas. We kept extending the garden until it went right around the house, with lawns, trees and flowers.

While bringing in the cows in the mornings we kept an eye open for field mushrooms, if we had no bag we tied them in our handkerchiefs. (I'm sure we always had clean hankies.) They were delicious fried with our morning eggs. If there was a glut of them, we went out with a bucket to pick them; Mum sometimes boiled them in milk for dinner. We bought very little food because we always had eggs, milk, vegetables and fruit, all kinds of meat – bacon, rabbit, hare, pheasant, partridge and poultry. In fact almost everything except bread, flour, tea and sugar.

September

September

September in the morn,
Black shapes rising all around,
The rooks seek every grain of corn,
Shadows on the stubble ground.

The sheaths of corn like statues stand,
In serried ranks and row on row,
The stubble shining silver gold,
In the early sunshine's glow.

The hedges deep with dewy flowers,
Meadow sweet and Withywind[¥],
Ivied oaks make shady bowers,
Around the fields at harvest time.

The golden hiles[§] are tipping over,
Weighed down with heavy ears of grain,
Time to take them under cover,
The tractors chugging up the lane.

[¥] Withywind or withywine, Dorset name for bindweed.

[§] Hiles, Dorset name for stooks.

September

September starts as other years, but our peaceful lives would soon be changing. The third of the month fell on a Sunday and Mum and I were at church as usual. On the way home we were cycling along by the council houses when people came out to tell us war had started, they must have heard it on the twelve o'clock news. Everyone felt a bit apprehensive, but nothing much happened for quite a while. A lot of the young men of the village were called up for the services and people were singing, 'We're going to hang the washing on the Siegfried Line, have you any dirty washing mother dear?', little thinking how quickly the Siegfried and Maginot Lines would be of little consequence. We had to be careful with lights, and everyone was making blackout curtains for all windows. We were still using candles at the time, but not even one little chink of light must show.

We had no wireless set at home, so one was very quickly bought, it was run on heavy batteries, and they had to be taken a few miles on the bicycle every now and then to be charged, we had to be careful that the acid did not splash out on our clothes. It was great to be able to hear the latest news, although everything was censored, and to hear the latest songs of the day, *Run Rabbit Run*, and *Roll Out the Barrel*. It was not long before Billy Cotton's band was playing for Workers' Playtime, with such songs as, *There'll Always Be An England, The White Cliffs of Dover, The Quartermaster's Stores* and a bit later, *Coming in on a Wing and a Prayer* and, of course, Vera Lynn and *We'll Meet Again*.

It was an exciting time for the young people, the boys were coming home on leave in their new uniforms but there was not much fighting going on for a while, it was like the lull before the storm. It was not long before everyone was issued with gasmasks; thank goodness we never had to use them. We were supposed to take them everywhere with us, they were in cardboard boxes about six inches square, inside little canvas carrying cases with straps. We were also issued with identity cards.

The farming went on much as usual, but the farmers soon had to start ploughing up more land and sowing corn, we ploughed up a few acres and sowed winter wheat. The blackberries were

plentiful in the hedges; we picked a lot for jam and also for wine. They were especially nice on the homeground and cowground road hedges, really sweet and firm, and warm from the sun; they smelled lovely in the baskets.

Michaelmas daisies started coming out in the garden, very tall and large-flowered, I think they were called Climax. The harvest festival was always celebrated in September. The church was decorated with fruit, vegetables and flowers and in the aisle near the chapel steps were always sheaths of corn. I loved the harvest hymns the best:

> We plough the fields and scatter
> The good seed on the land,
> But it is fed and watered
> By God's Almighty Hand.

also,

> Come, ye thankful people, come,
> Raise the song of Harvest-home,
> All is safely gathered in,
> Ere the winter-storms begin.

People loading up their trolleys in supermarkets today can have no inkling of what a good harvest meant to country folk at that time. There were no combine harvesters to get over the work quickly, and not many tractors to haul loads long distances, most farms were almost totally self-supporting. The shops were not full of foreign produce except for oranges and bananas, dried fruit and a few luxury items, although we did have the French onion sellers, riding their bicycles around the roads selling their strings of onions. The war put an end to many of the old ways. Things were never the same again. I suppose the changes would have come anyway, but the war accelerated it.

The apples were ripening in the orchard. Oh! What a lovely job, climbing the ladder into the trees laden with fruit, on sunny autumn days. The orchard must have covered about two acres, not all of it trees. It was bounded by the brook on one side, with oak and ash trees on the far bank. There were about a dozen apple trees, two were what we called Prophets, large juicy golden apples which kept until about Christmas, there were about three Cadburys, which were not so sweet but good keepers. There was a Russet and a small red apple which was red all the way through to the core. There were several others of unknown origin but the loveliest of all was the Blenheim Orange. It was large tree on which a swing had hung since I was a little girl. The fruit was a gorgeous orange colour and the taste was really out of this world. I've never tasted an apple like it since. We wore a special apron to pick the apples into, so we had both hands free in the tree, we then filled the clothes basket, and carried it into the house and up the back stairs to the cheese room. We spread newspapers on the floor around the walls, and piled the apples onto it. There was always a lovely appley smell and they kept us going in tarts, dumplings and apple cakes almost until apples came again. The apple cakes were a Dorset speciality; we added chopped apples, treacle and a pinch of spice instead of the dried fruit to the cake mixture.

The swallows are beginning to gather on the rooftops and on the top iron railing of the fence, they will soon be away.

October

Swallow Tails

Swallowtail suits of navy blue,
White shirt fronts, a uniform crew,
They came from nests in rafters high,
And launched into the wide blue sky.

Weaving, soaring, diving, darting,
Twisting, twittering, meeting, parting,
Line upon line on electric wires,
Ready to go, all expert fliers.

Preening, stretching, wings and tails,
Dipping, swinging, swallow tails.
Plumped up by England's summer store,
Midges and butterflies galore.

Soon they will leave this land of green,
In warmer climes they will be seen,
How many summers left to me,
To see them winging from the sea?

October

The swallows are gone, we miss their songs and their swooping and diving around the yard, we know the summer is gone and wonder what will happen before another summer although, so far, the war does not seem to be happening at all.

The leaves are beginning to turn colour, we will soon be back to winter routines again. Already the cows begin to stand around the yard gate, waiting to get in for a few tasty bites.

We were busy digging spuds. That was one farm job I did not enjoy, I was often left to pick them up on my own. Dad was still digging them by hand at that time. He probably didn't like picking them up either, as we got a considerable amount dug and then he would disappear to market, or some other place, leaving me in the middle of a field surrounded by potatoes. It was a rather soul-destroying job, and lonely too. Sometimes they were put in the barn and sometimes in a pit outside, as were the mangolds. We had quite a large house for mangolds, with thick stone walls which were usually lined with straw. When it was full up we made a pit with the rest. They were put in a long, large, heap then covered with straw or bracken with a thick layer of soil all over them, with a few little straw chimneys at the top for ventilation. We rented some land from the poultry farm up the lane; most of the mangolds were grown there. We usually pulled them about the third week in October, starting early in the morning. Sometimes the leaves were white with frost and our hands were frozen until we got used to it, then they started to burn and pain for while. We usually pulled enough to keep us going all day, laying them in double rows, leaves outside, then we went along each side with a sharp chopping hook cutting the leaves off very quickly. We loaded the mangolds on to the Overland and hauled them home. Some of them had grown to a huge size and it was hard work lifting them up. It was very easy to get slivers of mangold skin under the fingernails, which became very painful. They were very heavy loads for the poor old Overland, the wooden spokes slopped and creaked and almost came out of their grooves, though they never actually did. If the weather kept dry the job was soon done. Any mangolds which were left at the end of the day had to be carefully covered up with leaves in case of frost as the part which was in the ground would freeze and be ruined.

The pigs had the run of the farm to eat the acorns, of which there were plenty. Oak trees grew in all the field hedges and some of them were very fine. There were also a few across the middle of some of the fields, maybe there had been hedges there too at an earlier time. The hedges contained lots of hazel bushes. What could be nicer on a golden October day than to take a bag and go nutting? The two dogs always went too, they usually turned up a few pheasants and partridge from the hedgerows and also chased a few rabbits; there were rabbit burrows in every

hedge. Dad liked a few rabbits around the place, he never tried to get rid of them all completely. I saw a strange sight in the orchard one day, a rabbit fighting off a fox, and she won! The fox must have been as surprised as I was, because he took to his heels and ran.

For quite a few years after leaving school, I got no payment for the work I did at home, so I was always on the lookout for ways to make a few bob. One of the ways was mole catching. Moles were always called 'wants' at home and molehills were called 'want heaves.' The moles did a lot of damage in the fields, especially in the arable land, where mangold and swede seedlings were sometimes lifted out of the ground. Dad had quite a few mole traps, the oldest ones had wire spikes which speared the poor unfortunate animals but we never used those. I would set the traps in likely runs, the best ones were the ones coming out from the hedge, but these were deeper and had to be located by pushing a stick down into the ground. I made visits to the traps twice a day, they were often set off, and the moles escaped capture. The ones which were caught were nearly always dead, as the traps snapped together very tight, and the moles must have died almost instantly. Every day's catch was skinned and nailed out in neat oblong shapes on a board, or the back of a door. When they were really dry they were taken off and posted in neat shaped parcels to Horace Friend of

Wisbech. The skins were lovely soft fur, so thick and shiny, and I think I got about tuppence (two pence in old money) each for them.

Dad had many Dorset words: Basket to hold half cwt – *Bushel-basket*, Pitchfork prong – either *two grained, three grained* or *four grained*. Other words were: stay – *bide*; little girls – *maidens*.

November

November

The grey mist is hanging,
Its blanket around,
The trees stand in silence,
And birds make no sound.

    The leaves fall like raindrops,
    Of orange and gold,
    The cast offs of summer,
    Will crumble to mould.

        We're like falling leaves,
        So fair in the spring,
        We wear our bright dresses,
        We dance and we sing.

            The days of November,
            Will come to us all,
            We'll wither and perish,
            As leaves in the fall.

November

No sun, no moon, no morn, no noon. November was often the time for heavy fogs. Dad hated the fogs, he could not see the signs he was so endlessly looking out for. He watched the clouds, the sky, the moon, the winds, the animals, the birds, everything told him a story. He was a very accurate weather forecaster, the direction of the wind at new moon was very important to him. 'Twill be no better while this moon is in,' he would say. He would not sow seeds until the moon was in a favourable position. He knew if it was to be a hard winter by the numbers of migrant birds and whether they came early or late. He also knew the whereabouts of any foxes by the magpies chatter and the jays' shrieks, or by the crows' noise, and so was able to protect the poultry accordingly.

On clear days we had lovely views from the farm; to the south-east we could see Bulbarrow and the Dorchester Gap and to the north-east was Melbury Hill, like an enormous lion lying there, watching over us, its tail in Compton and its head at the beacon, with the furze bushes (fuzz bushes Dad used to call them) for a mane. The hill, which was over 860 feet, was composed of chalk; I often took home a lump to try and carve something out of it, as it was soft enough to chisel by hand, but I was never successful. At that time the chalk pit, which went about a third of the way up the hill, was a sparkling white patch which seemed to enhance its beauty. It could be seen for miles, but now it is covered by small trees, mostly ash. There was an old lime kiln at the bottom, in which we often sheltered from the rain. I can remember once Dad found a dead donkey in there and we had to go view it, as Dad said it was a rare sight to see a dead donkey.

There was a public footpath from the main road running near the chalk pit to the top of the hill, but the last time I was there, I'm sorry to say, the iron gates were padlocked, the paths were overgrown and there was barbed wire to stop people going to our beautiful hill.

With the coming of the colder weather we had to gather wood for the fires; all fallen trees and boughs were taken home. The crosscut saw was taken out of the shed. The saw must have been over five feet long, with a handle at each end, Dad and I took a handle each and we quickly got a good pile done.

The November evenings were getting so short, the lanterns were used once again for the evening milking. The evenings were spent by the cosy log fires; we had no electricity then. The table stood in the middle of the kitchen floor, and over it hung a white-shaded, well-polished, shining, brass paraffin lamp. One of the dogs always stayed indoors. Dear Wendy, my childhood companion, always slept on the sofa which, over the years, had gradually sunk lower and lower, until there was a deep dent. She was a very small greyhound, a dark brindle in colour. I remember the night my father fetched her as a puppy from Motcombe; one of those nights when he really did, 'Go to see a man about a dog.' This was always his answer when asked where he was going. The other dog stayed in a shed outside, her name was Ouida. Poor Wendy was getting old at this time. They were good dogs, they could run at great speed and catch anything they were put after – rabbits, hares, foxes, anything, but they never went off on their own.

I cannot kill anything now, not even a spider, but in those days I knew no other life, and the rabbits could not be allowed to become such a great plague. We used to hang the rabbits up in pairs by their back legs over ropes attached onto the beams in what was the former dairy. There were often sixty to one hundred ready for collection. The highest price I remember was four and a half old pence each. The rabbits were all scrutinised and maybe some would get less than others. Of course, that was only a winter occupation. Dad's brother Hubert usually came with us on our rabbiting expeditions. When we reached the chosen place, each hole was carefully netted, with the drawstring of the net hitched around a stick or stone. The ferrets were then put in; the dogs were kept on the leash. Most of the rabbits were caught in the nets, but escapees were shot at and the ones which escaped all that, were set after by the dogs. They flew like the wind and usually brought the rabbits back. Sometimes we had to wait ages outside the holes in the cold frosty weather, the rabbits would not bolt, and sometimes the ferrets killed them under the ground and we had to dig them out. Some of the ferrets were known as line ferrets; they had to go in on a long line or cord, as they were prone to kill the rabbits underground. The ferrets often had their long teeth shortened with pliers but this did not hurt them as it was only the tips of the teeth. If we did lose a ferret and all else failed, a nice warm nest of hay was put in the hole and all the bolt holes were stopped up. First thing in the morning someone had to go and see if the ferret was in the nest, and catch it if possible. Sometimes it took days to find them. Some of the ferrets were white and some were polecats. Waiting outside of rabbit holes on wet wintry days was a very cold job, Dad sometimes took pity on me and rolled me a cigarette to cheer me up. The rabbiting scene was carried on into the winter also, on cold, stormy nights we were often occupied with net making. Dad would cut out a flat piece of wood about a foot long, and maybe one and a half inches wide, according to the size of the mesh needed and with a smaller piece of wood for the bobbin, we would work away. Dad could make a couple of nets in an evening, I was slower. He also made rabbit wires, twisting several pieces of wire together, fixing the loop to strong twine, which was then tied to a wooden peg. The fox wires were made from old vehicle clutch wires, which he would get from the garage man.

Many evenings were spent skinning foxes and badgers, which were making away with the rabbits in large numbers. I used to dread them getting caught, though Dad was always as humane as possible, making constant journeys to look at wires, and putting the poor things out of their misery as quickly as possible. On moonlight nights we were often out with the rabbits. I used to dread the summons, happily warming my toes by the fire, 'Come on Tudge, let's go and see if the foxes are at the wires.' Dad always called me Tudge, in fact I never heard him call me any other name, I don't know where he got it from, most of my uncles called me that too, but Mum never, I was always Edna to her. When we were out in the moonlight I began to enjoy it, the fields and hills were so beautiful, and I hoped there would be no foxes that night. We used to stop under a large  oak tree and look down over the fields where the wires were, Dad would do his imitation of a rabbit screaming, which was quite realistic. He rolled his sleeve up above his elbow, and sucked on the bare skin in such a way as to make a loud scream. The foxes were completely taken in by it, and it wasn't long before a dark shadow would emerge from the copse at the far end of the field. We had to stand there absolutely motionless until the fox was within firing distance, and then bang.

I never thought there was anything odd about my lifestyle as I had never known any other. I was taught to move like an Indian hunter, if I as much as stepped on a twig I was severely reprimanded, there had to be absolute silence. It brought me very close to the world of nature, for which I have always been grateful.

December

Gales

*Gales in the woodland,
Twisting tipsy trees,
Dancing like dervishes,
Stripped of their leaves.*

*Gales in the big oaks,
Down by the brook,
Breaking the branches,
Disturbing the rook.*

*Gales round the farmyard
Are snatching the hay,
Blowing the animals'
Coats every way.*

*Gales around the garden,
Tossing the bushes,
Fluffing the feathers
Of robins and thrushes.*

*Gales pound the farmhouse
And howl at the gable,
Rattling the doorways,
To get in – unable.*

December

The short dark days of winter came around again, it was lovely to sit around the fire on wintry nights and listen to the gales blowing outside. The farmhouse walls were built with cut stone about two feet thick; it was always warm and cosy inside. The oldest part of the building was probably over two hundred years old.

The sitting room windows had wooden shutters on the inside to keep out the cold; they had hinges in the middle so they folded up small enough to fit into the sides of the windows. The kitchen, as the living room was called, contained a comfortable three piece suite, creamy grey, with a small floral pattern, which my mother had bought from a mail order firm, I think it was Pontings, when she married in 1923. She paid two pounds, fifty shillings for it, and she also received as a free gift, an oak gate-legged table which is still in use today. Also in the kitchen was a sideboard, which had started life as a spinet. There were five bedrooms in the house, one of which was the huge cheese room. In one room was a five sided window near the ceiling, it always reminded me of, 'The little window where the sun came peeping in at morn,' of the *I Remember* poem. Each bedroom had its marble topped washstand, with china wash basin, large jug, soap dish and toothbrush holder, under each bed was a Jerry (chamber pot) to match. My mother's bedroom had a most beautiful set, it was black china with life-size peaches which looked just like real ones, the peach colour and green leaves showed up to perfection on the black. We used earthenware hot water bottles; the warming pan hung on the wall but it was gone out of use by that time. It was such a lovely old rambling house, there were steps leading up to most of the rooms and bedrooms as the house rambled its way up the slope. How lucky I was to live in it.

The rabbiting was in full swing again with the cold weather, we hurried to get all the jobs finished for an early start. As the milk lorry came so early, about eight o'clock, there was no excuse for a lie in any day. Someone always had to give the lorry driver a lift up with the churns; it was quite an art swinging them up and not getting our fingers trapped by the edge of the churn. We had one very nice lorry driver, we called him Clark Gable as he looked exactly like the American film actor. One of the drivers was not nice at all, when I told Clark Gable about him he soon sorted him out, and the next time the 'baddie' came he actually apologised for his behaviour. After the milk was gone we quickly did all the feeding and cleaning out, the cows were still outdoors in the daytime, with any luck they stayed out until January. If the ground became too wet we had to bring them inside earlier. Before Christmas came we got in extra loads of hay to keep the cows going for the holiday.

On Christmas morning we always attended church and the building rang with the sound of carols, *'Oh Come All Ye Faithful,' 'While Shepherds Watched Their Flocks By Night,' 'Holy Night,'* etc. We were watched over by the lovely stone angels at the base of the roof supports. We always spent the rest of the day at Granny's house at Compton, all the family gathered there. The uncles had not yet been called up for military service, as they were in their thirties, so things were much the same as other years, except for the blackout. Christmas was celebrated in great style, home grown hams, turkey, wines, cream, puddings and cakes, etc. crackers by every plate, followed by songs such as, *Red Sails in the Sunset, Little Old Lady, The Isle of Capri, Roll Out the Barrel, Horsey, Horsey*, and I think the *Palais Glide*, was out then. Granny sang a funny song which began, *There was an old farmer had an old sow*. There were games such as Consequences and Black Magic, followed by cards; the favourite game seemed to be Newmarket. There was a break about four o'clock in the afternoon, when the farmers returned home for afternoon milking. We usually rode our bikes; it was a distance of about two miles from our house. On the way home that dark night, with the lights dimmed for the blackout, I rode into a black cow which was lying asleep in the middle of the road, not much harm was done though, once I realised what sort of animal I had collided with.

Christmas Day was really the only holiday; Boxing Day was the big rabbiting day of the year, when we were joined by relatives eager for a day out. We usually started out about ten o'clock. After assembling the ferrets, dogs, guns, spades and all the usual paraphernalia, we made for a field called Brockly, belonging to Willis Farm where two of the uncles lived. We proceeded to net all the holes, being careful to find the bolt hole which was usually well hidden behind the bank. When all was ready the ferrets were put in, usually two of them, and we soon heard rumbling noises coming from under the ground. Then out popped rabbits into the nets, the nets all had drawstrings on them fastened to a bush, and the poor rabbits could not escape. Some refused to come out and the ferrets ended up killing them, which meant we had to dig them out. Some always escaped the nets and were usually shot or caught by the dogs. Sometimes to make more sport on Boxing Day, no nets were put down at all, it was left to the dogs and guns. It was often the only day of the year that all the brothers were together and there was much 'chaffing' going on. I was happy to be in charge of the ferrets, most of them were likeable little animals, especially the white ones, with their little pink eyes, although I was careful to pick them up by the back of their necks, just in case! This day ended up with a big happy party at my Aunt Cis's house.**

And so the year drew to a close; nothing would ever be the same again. We did not know how the war would go but it sometimes seemed as if it had not actually started. We just kept on with our everyday jobs, with just a little apprehension, we were all sure we would win in the end; and we did!

** Now the Milestones Tea Rooms adjacent to Compton Abbas Church.

Tree-felling

Edna Rice (nee Hunt)

Compton School about 1930. Edna is in the 2nd row from front, direct centre.

Compton to Melbury road

Uncle Joe Miles

Edna's cousins

Edna and cousins

Edna in her 'halo' hat

Aunt Cis

Wuffie with Edna's mother

Home-guard

Edna with Wendy and Ouida